Air Force Research Laboratory 2014 Strategic Plan

Purpose

The 2014 Air Force Research Laboratory (AFRL) Strategic Plan will shape and guide AFRL's actions for the next 3 to 5 years. Shaped by higher headquarters' guidance and direction, particularly the *Air Force Science and Technology (S&T) Strategy, Technology Horizons: A Vision for Air Force Science & Technology during 2010-2030*, and *Global Horizons Final Report: United States Air Force Global Science and Technology Vision*, this plan will serve to focus organizational resources to ensure that AFRL will continue to provide the right technology, at the right time, and at an affordable cost.

Communicating the Strategic Plan

Directors and leaders at all levels should make this plan available to their people and be prepared to explain its contents and principles.

Air Force Research Laboratory 2014 Strategic Plan

Science & Technology Strategy Summary

Global Environment

The world is experiencing an era of rapid change and globalization. Increased competition for resources, access to Information Technology (IT), and changing demographics has the potential to shift the balance of power. The Air Force is facing conditions that diverge significantly from the strategic environment of the last two decades. Potential adversaries are using emergent globalized technology and manufacturing infrastructure to rapidly develop sophisticated military capabilities that create more contested operational environments. The challenge is to ensure our Airmen obtain the best technology, at the right time, while affordably meeting mission needs.

S&T Guiding Principles

The Air Force is committed to a strong S&T program that will enable fully integrated air, space, and cyberspace forces meet the challenges of the 21st Century. Our S&T program lays the technological foundation for the current and future Air Force to assure America's security through *Global Vigilance, Global Reach, and Global Power*. The cornerstone of S&T's innovative power is its exceptional resources: S&T experts and our unique equipment and facilities. The Air Force, through AFRL, manages Air Force S&T as an integrated program by using its special resources to invest in future capabilities a n d provide the warfighter near-term technical support. AFRL carefully balances the investment portfolio in basic research, applied research, and advanced technology development, allocated between in-house and contracted activities, to produce both evolutionary and revolutionary technologies focused on Air Force service core functions and capabilities.

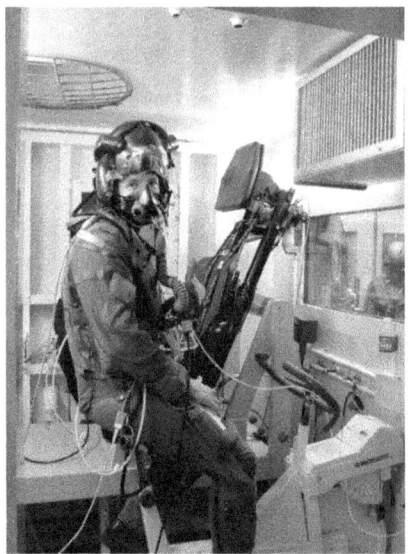

Air Force S&T Guiding Principles

- Address the highest priority needs of the Air Force across the near-, mid-, and far-term
- Execute a balanced, integrated S&T Program that is responsive to the Air Force Core Missions
- Advance critical technical competencies needed to address the full range of product and support capabilities

Air Force Research Laboratory 2014 Strategic Plan

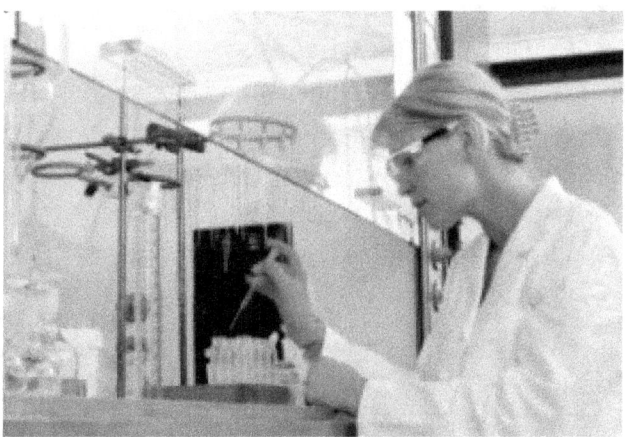

In addition to the S&T Guiding Principles, and based on the current strategic, fiscal, and operational environment, the Secretary of the Air Force has given the Air Force S&T Program seven specific S&T Goals which serve to focus the available resources on issues of critical importance to the Air Force. These goals should receive increased emphasis and serve as a guidepost for resource decisions.

They are as follows:

Air Force S&T Strategic Goals

- Leverage and create technology trade space with sufficient technical and manufacturing maturity to support near-, mid-, and far-term acquisition programs
- Innovate technical solutions to rapidly respond to urgent warfighter needs
- Develop concepts and create new science and technology options that address threats and maintain or increase capability, readiness, and availability at reduced cost
- Invent concepts and science and technology options that enable new missions or new capabilities to support *Global Vigilance, Global Reach, and Global Power*
- Employ business practices that increase the S&T Enterprise's inventiveness, productivity, and responsiveness to Air Force needs
- Acquire, develop, and retain a talented and high performing S&T workforce
- Invest in core S&T infrastructure to ensure its future health

3

Air Force Research Laboratory 2014 Strategic Plan

AFRL Overview

Headquartered at Wright-Patterson Air Force Base (AFB) OH, AFRL leads a worldwide government, industry, and academia partnership in the discovery, development, and integration of affordable warfighting technologies for the Air Force. AFRL is the single Air Force laboratory, responsible for planning and executing the entire Air Force S&T program with world-class facilities nation-wide. The laboratory provides leading-edge warfighting capabilities and revolutionary technologies that keep our air, space, and cyberspace forces the world's best.

VISION – "We defend America by unleashing the power of innovative air, space, and cyberspace technology."

Since its inception over 65 years ago, the United States Air Force has continually endeavored to be the most technologically advanced Air Force in the world and it has succeeded. To maintain that advantage, it is crucial for AFRL to continue providing leading-edge and innovative S&T. The imagination of AFRL's highly-skilled and technical workforce will continue to unleash the power of air, space, and cyberspace and will do so far into the future.

MISSION – "Leading the discovery, development and integration of affordable warfighting technologies for our air, space, and cyberspace forces."

AFRL's mission is unchanged. AFRL must remain on the forefront of efforts to advance the state-of-the-art, to fully develop new and affordable technologies, and to integrate technologies so as to ensure our Air Force is equipped to defend our Nation.

Air Force Research Laboratory 2014 Strategic Plan

AFRL Headquarters

AFRL Headquarters staff provides the workforce and infrastructure necessary to ensure that AFRL can accomplish its mission. The headquarters staff assists the Commander to formulate and disseminate policies, plans, and directives affecting the lab and ensures execution of all HQ AFRL business responsibilities.

Director of Staff (AFRL/DS)

The Director of Staff is the Commander's primary interface with the staff for the daily operations of the Headquarters. The Director of Staff manages and integrates all activities and facilities of the Center Headquarters. Through the Commander's Action Group and the Integration & Operations Division, the Director of Staff provides executive support and services, manages the center's tasking process and provides IT, communication and facility support for the Headquarters' staff. The Director of Staff is responsible for center-wide security and logistics policy. The Director of Staff manages the suspense control process (SOCCER) for AFRL. The Director of Staff manages the AFRL History Office and through the Operations Support Division provides support to the Headquarters and Directorate military personnel. The Director of Staff is the center's primary liaison to the installation public affairs, protocol, inspector general (complaint resolution) and safety offices.

Operations, Test & Evaluation (AFRL/DO)

AFRL Operations develops and provides technical directorates policies, processes, and tools for research test activities to include related ground, flight, and space operations. DO supports AFRL program managers and scientists & engineers in research test matters including the development and execution of in-house and contractor test plans. DO is responsible for all test management and research operations support functions related to research test activities and serves as the interface to AFMC and Air Force test and evaluation policy makers. DO provides flight operations oversight and executes authority for AFRL small unmanned aerial system flight tests and operations. The operations office ensures compliance with all DoD and Federal Communications Commission regulations with respect to the certification and use of radio frequencies for existing and new technologies.

Safety (AFRL/SE)

AFRL Safety provides advice to the AFRL Commander on the establishment of a proactive mishap prevention program to assist supervisors at all levels with their safety responsibilities. SE develops policy, administers and evaluates AFRL's flight, Voluntary Protection Program, ground, system, weapon, and space safety programs.

Air Force Research Laboratory 2014 Strategic Plan

Personnel Directorate (AFRL/DP)

AFRL/DP shapes the AFRL workforce and workplace by providing integrated resource management services. The organization supports over 5,000 military members, civilian employees, and laboratory leaders assigned to AFRL by developing policies and processes that foster recruitment, retention, and professional development. DP supports numerous recruitment activities throughout the country to promote AFRL as an employer of choice and to ensure the highest-quality individuals are made aware of the benefits of AFRL employment. DP provides support for developmental opportunities through professional development programs, professional military education, and leadership programs. The directorate is responsible for formulating human resources policy and guidance, and establishing and managing human resources processes in AFRL, affecting both military members and civilian employees.

Engineering Directorate (AFRL/EN)

AFRL/EN provides engineering and program management rigor and advice across the AFRL S&T portfolio to focus technology development on high priority needs and to transition technology to effective and affordable weapons systems. The organization instills systems engineering and programmatic discipline and rigor to focus limited S&T resources on the most important capability needs and the most promising technical solutions within cost and schedule constraints. EN establishes a clearly-defined airworthiness and flight test processes and ensures the workforce is trained on the process. The directorate provides program and investment management tools and processes that add value to the workforce, and assists in ensuring AFRL has a capable engineering workforce to meet Air Force capability needs now and in the future.

Financial Management Directorate (AFRL/FM)

AFRL/FM provides world-class financial services supporting AFRL operations; ensures responsible stewardship and public accountability of resources; and, provides decision makers with accurate and timely financial information. The organization establishes policy, oversees, and reports program execution; transmits uniform guidance and direction to all laboratory organizations; and, orchestrates AFRL budget execution and resource drills with participation from across the laboratory. FM provides expert cost, economic and financial decision support through the use of Activity Based Costing, Business Case Analysis, Economic Analysis, Cost Estimates, Earned Value Management, and other cost tools. FM ensures financial standardization and integration of business information management in support of AFRL operations, and validates and analyzes AFRL compliance with established accounting principles and regulations.

Air Force Research Laboratory 2014 Strategic Plan

Contracting Directorate (AFRL/PK)

AFRL/PK's role is to position AFRL's contracting workforce and operations to attain leading-edge scientific capabilities and technology transfer through innovative contracting techniques, facilitation of small business research and robust workforce development. PK's management strategy objectives focus on enabling seamless operations across the five contracting Geographically Separated Units to achieve the PK corporate goals. PK objectives include enhancing AFRL's ability to accomplish its mission by institutionalizing best practices and fostering continuous process improvement, creating standardized corporate processes that enhance AFRL's infrastructure in order to increase organizational productivity, reduce administrative burdens, and maximize the effectiveness of our resources.

Research Collaboration & Computing Directorate (AFRL/RC)

AFRL/RC enables mission success by providing world-class, high performance computational resources, enterprise-wide business intelligence and information management systems, and information technology strategic planning and direction. The Directorate provides these capabilities through three operational divisions and one support division. The Supercomputing Resource Center, as part of the DoD High Performance Computing Modernization Program, provides high performance computing platforms, high-speed computer networks, and classified vault operations. The Enterprise Business Systems Division develops, deploys and sustains enterprise IT solutions to support AFRL business processes. The Information Strategy and Policy Division develops AFRL corporate investment strategies, plans, and direction to implement Air Force IT and Information Assurance policies. The Integration and Operations Division provides the directorate with a wide range of support functions including contract, financial and workflow management.

Small Business Directorate (AFRL/SB)

The mission of the AFRL Small Business Office (AFRL/SB) is to maximize opportunities for small businesses to deliver technology innovations and solutions to meet customer needs. SB's duties include aiding, assisting and counseling small businesses, assisting in formulation of acquisition strategies, and to take the role of ombudsman for small business issues. The SB team is dedicated to creating and delivering strategies that bring innovative, agile and efficient small business solutions to our customers in AFRL. The priority of SB is to deliver the right small business options and solutions to our customers, increase awareness of small business capabilities and their contributions to the AFRL mission, and educate internal and external audiences on meeting AFRL's mission with small business solutions.

AFRL/XP is responsible for the planning and programming of S&T resources to meet the warfighter's needs. The organization is responsible for design and execution of the AFRL Programming, Planning, Budgeting and Execution process; Congressional interactions; enterprise business IT requirements; and serves as the focal point for efficiency initiative reporting. The organization enables domestic S&T partnerships/alliances and international cooperations that leverage external resources to meet Air Force technology needs. XP leads the management and execution of customer engagement and represents stakeholder views to AFRL leadership. XP is the focal point for integration and coordination of special program S&T activities, and provides intelligence support for AFRL efforts. The directorate leads AFRL strategic planning and transformation through the development and management of enterprise level guidance and processes. XP is responsible for the development of this plan and implementing the metrics that will show progress towards meeting the plan's goals and objectives.

Air Force Research Laboratory 2014 Strategic Plan

Technology Directorates

AFRL is comprised of nine Technology Directorates (TDs) sites across the nation with major locations at Wright-Patterson AFB OH; Eglin AFB FL; Kirtland AFB NM; Rome NY; Edwards AFB CA; and Arlington VA. Each TD focuses on a particular technology area, but uses cross-TD collaboration whenever possible to combine technologies and demonstrate capabilities to higher technology and manufacturing readiness levels.

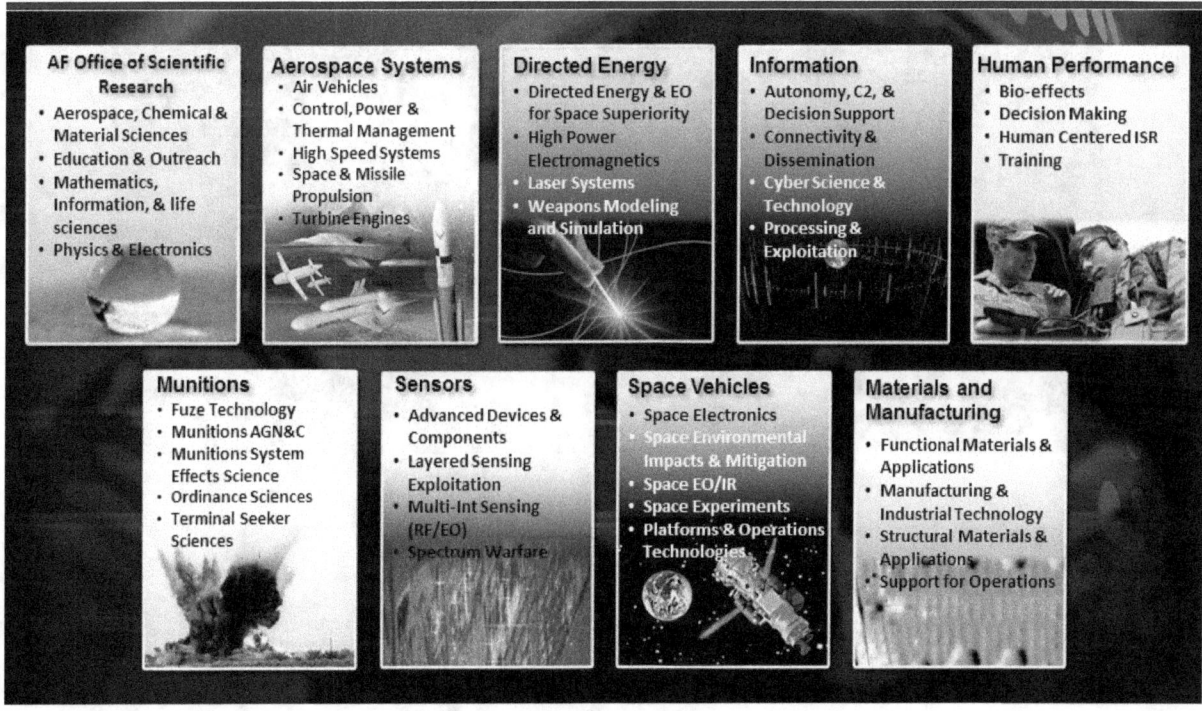

AGN&C – Aerodynamics, Guidance, Navigation & Control
C2 – Command & Control
EO – Electro Optical
IR – Infrared
ISR – Intelligence, Surveillance, and Reconnaissance
RF – Radio Frequency

Air Force Research Laboratory 2014 Strategic Plan

The Air Force Office of Scientific Research, located in Arlington VA, provides leadership and management of the AFRL basic research program. AFOSR also maintains overseas offices in London, Tokyo, and Santiago to provide the Air Force access to leading international research. AFOSR's mission is to discover, shape, and champion basic science that profoundly impacts the future Air Force. AFOSR invests in long-term, broad-based research in science and engineering. The majority of the research is conducted within United States academia, AFRL, and industry; a modest fraction is conducted by foreign academic institutions and research laboratories. The investment is distributed among more than 300 academic institutions, 150 industrial laboratories, and 260 AFRL research efforts.

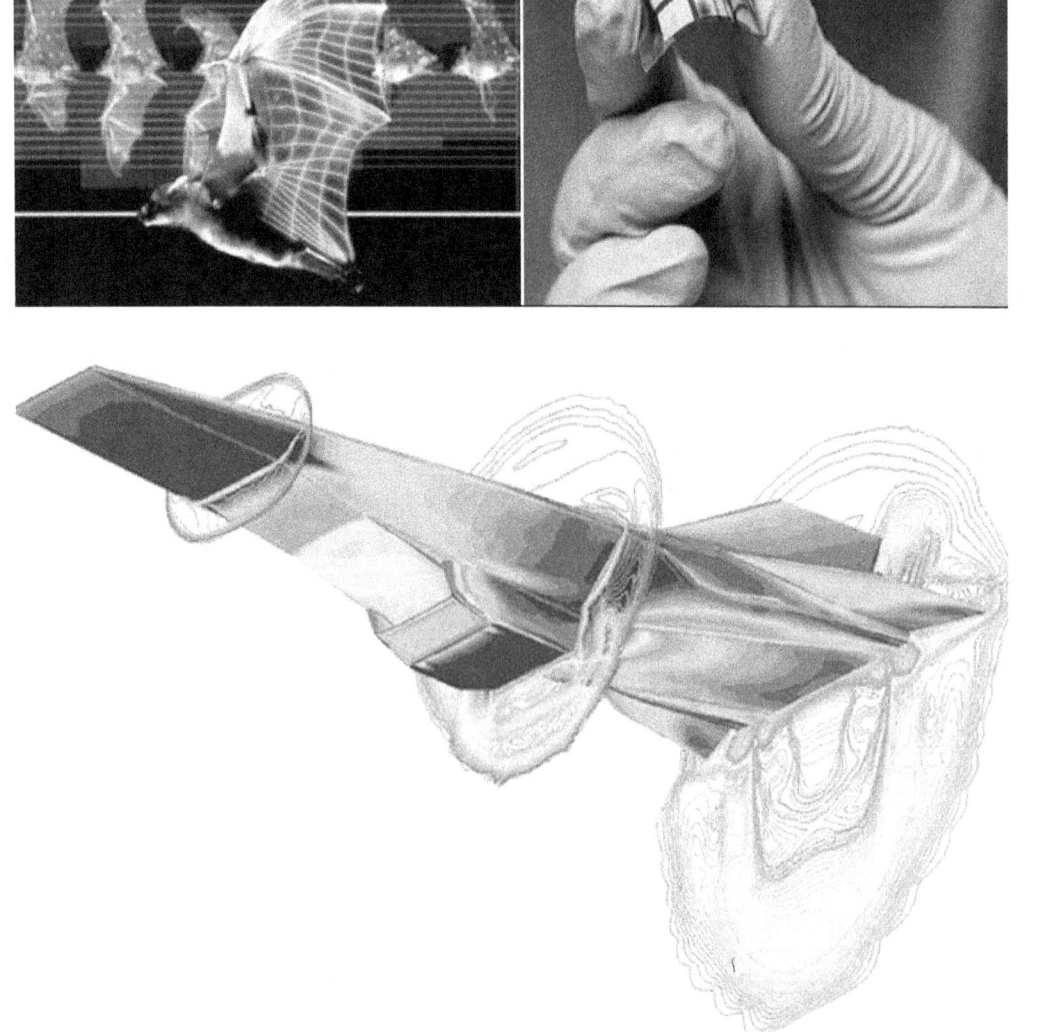

Air Force Research Laboratory 2014 Strategic Plan

Aerospace Systems Directorate (AFRL/RQ)

The Aerospace Systems Directorate leads the discovery and development of world-class S&T of integrated Aerospace Systems for national security. Established in July 2012, this TD brings together the former Air Vehicles and Propulsion Directorates and is based at Wright-Patterson AFB OH with an additional research facility at Edwards AFB CA. The organization's primary areas of focus include: Turbine Engines, Rocket Propulsion, High Speed Systems, Power and Control, Aerodynamics, and Structures. With a workforce of over 800 employees, the Aerospace Systems Directorate is responsible for the operation of several world-class facilities including: a fuels research facility, structural testing labs, compressor research facility, rocket testing facilities, supersonic and subsonic wind tunnels, flight simulation lab, and many other cutting-edge research capabilities. Building on a distinguished history of service to our nation, this organization creates the future of military aerospace.

The Directed Energy Directorate is located at Kirtland AFB NM, with a detachment in Maui HI. This TD is the Air Force's center of expertise for directed energy and optical technologies. The goal of this TD is to develop and transition technology in four core technical competencies: Laser Systems, High Power Electromagnetics, Weapons Modeling and Simulation, and Directed Energy and Electro-Optics for space superiority. AFRL pioneered the first and only megawatt class airborne laser and is a world leader in ground-based space imaging using adaptive optics employed in ground- based telescopes. Currently, this TD is developing game-changing counter-electronics weapon technologies that can degrade, damage, or destroy electronic systems with minimum collateral damage.

Air Force Research Laboratory 2014 Strategic Plan

Information Directorate (AFRL/RI)

The Information Directorate, located in Rome NY, provides S&T solutions to address critical warfighter needs and offer innovative game-changing options for the future United States Air Force in affordable Command, Control, Communications, Cyber, and Intelligence (C4I) capabilities. The C4I S&T program invests across a broad portfolio, attaining a balance between near-term, quick-reaction capability support; mid-term technology development to modernize the force; and revolutionary technologies that will provide future warfighting capabilities. AFRL/RI's robust portfolio is made possible by collaborating with and leveraging the resources and talents of other services and agencies across the Air Force, Department of Defense (DoD), Intel community, and other government, as well as industry and academia. The Information Directorate leads four main technical competency areas: 1) Autonomy, Command & Control Decision Support; 2) Cyber Science and Technology; 3) Communications and Dissemination; and, 4) Processing and Exploitation. RI has multiple state-of-the-art research facilities and is adding a Controllable Contested Environment to the Stockbridge Unmanned Aircraft Systems test site.

711th Human Performance Wing (711 HPW)

The 711th Human Performance Wing (HPW) is the first wing to consolidate human-centric warfare research, education, and consultation under a single organization. Established in March 2008 under the Air Force Research Laboratory and headquartered at Wright-Patterson AFB OH, with a detachment located at Fort Sam Houston TX, the 711 HPW is comprised of the Human Effectiveness Directorate, the United States Air Force School of Aerospace Medicine, and the Human Systems Integration Directorate. The 711 HPW's mission is to advance human performance and protection in air, space, and cyberspace through research, education, and consultation. The Wing supports the most critical Air Force resource – Airmen. The Wing's primary focus areas are aerospace medicine education/consultation/ research, human effectiveness S&T, and human systems integration. In conjunction with the Naval Medical Research Unit – Dayton and San Antonio, and national universities and medical institutions, the 711 HPW functions as a Joint DoD Center of Excellence for human performance sustainment and readiness, optimization, and enhancement.

Air Force Research Laboratory 2014 Strategic Plan

Human Effectiveness Directorate (711 HPW/RH)

Located at Wright-Patterson AFB OH and Joint Base San Antonio TX, the Human Effectiveness Directorate discovers biological and cognitive S&T to increase mission effectiveness. Research focuses on aerospace toxicity, cognitive and physiological performance challenges presented by next- generation platforms and their flight envelopes; challenges that confront cyber and command & control intelligence, surveillance, and reconnaissance (ISR) operators; and the challenges presented by world-wide growth in anti-access/area denial capabilities. Additional research areas include human performance augmentation, autonomy, decision making, trust, training, and directed energy bioeffects.

United States Air Force School of Aerospace Medicine (711 HPW/USAFSAM)

Located at Wright-Patterson AFB OH, the USAF School of Aerospace Medicine conducts initial and advanced officer and enlisted professional training for bioenvironmental engineers, public health, aerospace physiology and aerospace medicine. Research teams focus on human performance, en route care, expeditionary medicine, and force health protection. Epidemiology labs process more than 45,000 lab samples from around the globe weekly, conducting virology and bacteriology studies as a crucial part of our nation's epidemiological surveillance activity. Occupational & environmental health labs analyze 18,000 chemistry, 10,000 radioanalytic, and 14,000 dosimetry samples annually to enhance health protection for uniformed and civilian Air Force personnel.

Human Systems Integration Directorate (711 HPW/HP)

Located at Wright-Patterson AFB OH, the Human Systems Integration Directorate facilitates the application of human systems integration (HSI) principles to optimize Air Force operational capabilities at an affordable cost. Major areas of emphasis include consultative support to MAJCOMs and acquisition programs, HSI risk assessments at key acquisition decision points, and optimization of medical care delivery for Airmen and their families.

Located at Eglin AFB FL, the Munitions Directorate develops, and demonstrates S&T for air-launched munitions to defeat ground-fixed, ground-mobile/re-locatable, and airborne targets to assure pre-eminence of the United States Air Force. Focus areas include weapon guidance, navigation, and control; terminal seekers; damage mechanisms sciences; and, energetic materials and fuze technologies. Research and development ranging from basic research to advanced demonstrations is accomplished with a diverse workforce and world-class test ranges/facilities located at Eglin AFB. These research areas provide the Air Force with a strong revolutionary and evolutionary technology base upon which future air-delivered munitions can be developed to neutralize potential threats to the United States.

Air Force Research Laboratory 2014 Strategic Plan

Sensors Directorate (AFRL/RY)

Located at Wright-Patterson AFB OH, the Sensors Directorate develops new technologies that the United States warfighters need to find and precisely engage the enemy and eliminate its ability to hide or threaten our forces. In collaboration with other organizations, the program develops critical sensing and spectrum warfare technologies to enable *Global Vigilance, Global Reach, and Global Power*. Sensing and spectrum warfare technologies provide a complete and timely picture of the battlespace for global ISR, as well as battlespace awareness, targeting, and survivability for Air Superiority and Global Precision Attack. The main technology areas include; passive and active radio frequency, and electro-optical systems; precision navigation and timing (PNT); layered sensing exploitation; avionics vulnerability; and, cognitive, distributed spectrum dominance.

Air Force Research Laboratory 2014 Strategic Plan

Space Vehicles Directorate (AFRL/RV)

The Air Force's center of excellence for space technology is located at Kirtland AFB NM.   The Space Vehicles Directorate develops and demonstrates space technologies across the AFRL enterprise for more effective, more affordable space warfighter missions. Leveraging commercial, civil, and other government resources, the TD ensures the nation is always at the forefront of space technology. The TD performs basic research, applied research, and advanced technology development to understand and exploit the effects of the battlespace environment; develop improved/revolutionary technologies for space Electro-optic/infrared sensing; improve satellite payload and operations technologies; advance space electronics capable of surviving in the space environment; and, demonstrate these technologies for transition. This work directly supports the warfighter needs for space communication; PNT; space ISR; protection from both natural and manmade threats (Defensive Space Control); and, space situational awareness. It integrates the only Air Force S&T activity aimed at reducing cost of ownership of space systems while increasing performance and reliability towards enabling new capabilities.

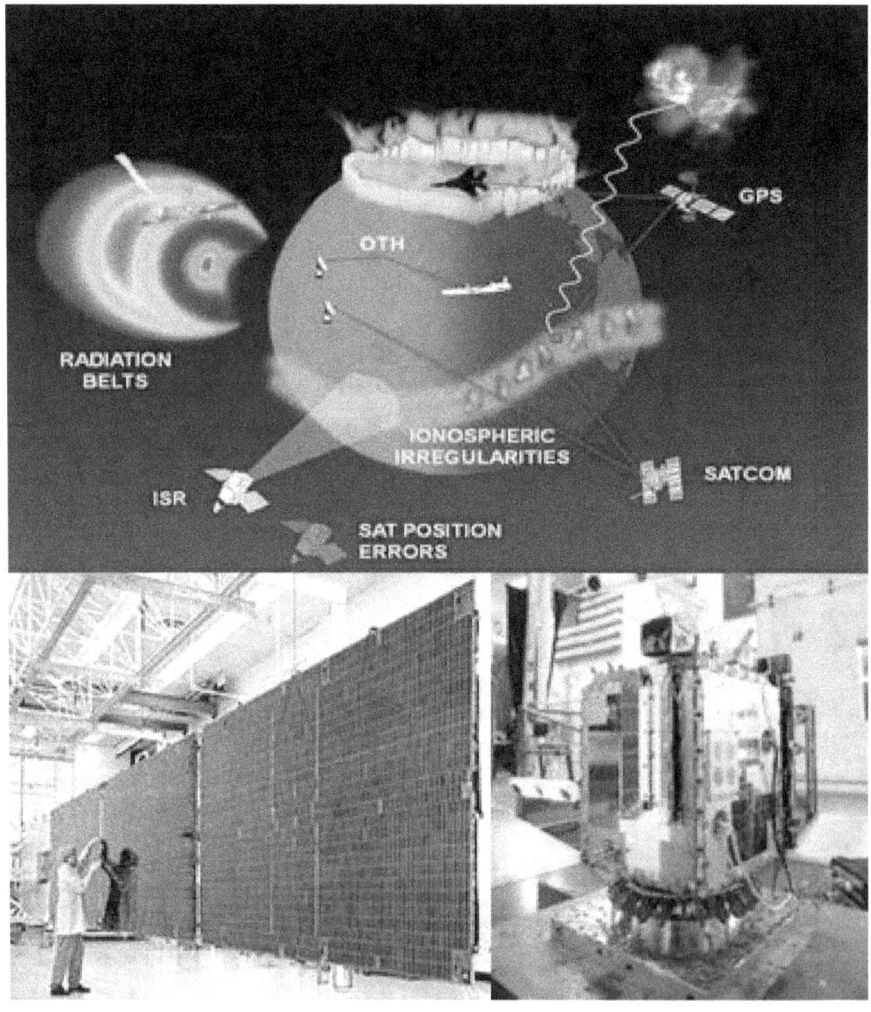

Air Force Research Laboratory 2014 Strategic Plan

Materials & Manufacturing Directorate (AFRL/RX)

The Materials and Manufacturing Directorate is situated at Wright-Patterson AFB OH and serves as the center of materials and manufacturing development for the affordability, sustainability, lethality, and survivability needs of current and future Air Force weapon systems. The TD maintains a full spectrum of exploratory and advanced materials research, and supports industrial preparedness by advancing manufacturing technology. Its dedicated and highly capable workforce conducts internationally recognized research that has advanced the state-of-the-art in a variety of technical fields including nano- and biotechnology, agile and additive manufacturing, nondestructive evaluation, innovative computational tools, as well as sustainment technologies for increased affordability and availability. This expertise, coupled with engagement through contractual partnerships with industry, ensures a range of robust technical options for future Air Force systems and for rapid response to urgent warfighter needs. The design and development of revolutionary materials systems and manufacturing processes provides solutions to the complex challenges of today and guides the vision of tomorrow.

In close coordination with the acquisition and requirements communities, AFRL will focus on reducing technical risk in current and future programs of record by delivering appropriately mature affordable technologies, tailored to customer needs. AFRL will use technology demonstrations and prototypes to prove new concepts, reduce integrated system S&T risks, create data for program decisions, or maintain key design capabilities in our industrial and manufacturing base.

AFRL will leverage outside resources to ensure Airmen receive the best technology to meet their needs. While our S&T investments are a prime source of innovation for the Service, domestic and international investments in S&T now dwarf our internal resources. AFRL will include efforts to guide and harvest S&T developed outside of our core investments and leverage them for warfighter use.

This Strategic Plan outlines the seven S&T Strategic Goals the Secretary of the Air Force has given the Air Force S&T Program and objectives for each goal. This guidance serves to focus the available resources on issues of critical importance to the Air Force.

S&T Strategic Goal 1: Leverage and create technology tradespace with sufficient technical and manufacturing maturity to support near-, mid-, and far-term acquisition programs.

In an era of fiscal constraint, it is important to create and transition technologies that meet the requirements of the warfighter community and respond to the needs of the acquisition community at the technical and manufacturing maturity level that is appropriate for the capability being developed. AFRL will achieve this goal by accomplishing the following objectives:

Objective 1.1: Increase transition of technology to warfighter systems.

AFRL will leverage all vehicles and processes available to us to more effectively transition technology to system acquirers, industry, and ultimately our warfighting customer.

Objective 1.2: Emphasize the use of demonstrations, prototypes, and virtual prototypes to mature technology, reduce acquisition risk, and create alternative solutions for acquisition programs.

Where possible and when it makes sense, AFRL will concentrate on reducing the risk associated with technology transition by employing demonstrations, prototypes, and virtual prototypes to achieve a more robust trial of technologies with the aim of reducing or eliminating unforeseen technical and manufacturing issues. AFRL will support this policy by looking for opportunities that promise to address capability gaps and create alternative solutions to surprise our potential adversaries.

Objective 1.3: Strengthen relationships with the acquisition, industrial, and warfighter communities.

AFRL will collaborate with the acquisition, industrial, and warfighter communities to pursue our S&T objectives. AFRL will identify technology needs, coordinate technology investment opportunities, and develop technology integration and transition strategies that leverage the work of others in order to achieve synergistic effects and produce "win-win" outcomes.

S&T Strategic Goal 2: Innovate technical solutions to rapidly respond to urgent warfighter needs.

Over the last decade, AFRL supported urgent warfighter needs through its Rapid Innovation Process and direct support from its world-class subject matter experts. AFRL's rapid support process capitalizes on AFRL's technical knowledge and innovative spirit; commercially-available technologies and systems; and, ready access to emerging capabilities that promptly deliver innovative solutions to the warfighters' most urgent needs. AFRL must retain and exercise the ability to quickly respond to urgent needs whether in peacetime or at war. To meet this goal, AFRL will accomplish the following objectives:

Objective 2.1: Reliably deliver rapid response solutions to urgent warfighter needs within the agreed-to time frame.

Providing technology solutions to the warfighter in order to meet time-critical needs is a hallmark of AFRL service. AFRL will strengthen its position as the "go-to" entity for solutions when our warfighter customers have urgent problems they cannot overcome without a materiel solution. AFRL solutions can range from the technical assessment required to keep an entire fleet airborne after a mishap, to providing rapid expert technical consultation, to delivering a new warfighting capability in short order. AFRL seeks to provide solutions that work and are delivered when promised.

Objective 2.2: Institutionalize and exercise wartime lessons learned and best practices to ensure AFRL retains its ability to rapidly respond to urgent warfighter needs.

AFRL will ensure lessons learned and best practices from recent technology deliveries are codified and exercised to enable us to deliver faster and more credibly into the future.

S&T Strategic Goal 3: Develop concepts and create new science and technology options that address threats and maintain or increase capability, readiness, and availability at reduced cost.

AFRL will consider affordability to include total life-cycle cost in all stages of S&T development. AFRL should understand the cost drivers in current capabilities and shape S&T research to find less costly technology-based solutions.

Objective 3.1: Create S&T options that affordably address Service Core Function gaps.

In addressing the Air Force's highest priority needs, AFRL will work with the warfighting, acquisition, and sustainment communities to understand the key cost drivers in customer mission and sustainment needs and seek technology alternatives that align with Core Function Support Plan gaps and reduce cost without capability decrements.

Objective 3.2: Identify and respond to weapon system cost drivers early and focus S&T initiatives to enable reduced life-cycle costs.

AFRL will provide focused technology efforts that lead to reduced acquisition and sustainment costs across the entire weapon system life cycle, including the operations and maintenance of those systems. Today, AFRL is building expertise in costing and cost analysis in all stages of the S&T acquisition process. AFRL will explore ways to expand its cost, economic, and financial decision support in this domain.

S&T Strategic Goal 4: Invent concepts and S&T that enable new missions or new capabilities to support *Global Vigilance, Global Reach, and Global Power.*

AFRL is always preparing for the future fight. However, the future will rarely reflect current requirements or identified capability gaps. In order to provide game-changing, new capabilities to the warfighter, AFRL must better understand existing and emerging military challenges and continue to search for and invest in disruptive and enabling technologies. Fundamental research should be geared toward exploring and expanding beyond the current state-of-the-art capabilities, but will remain focused on creating value for the Air Force. To keep the Air Force dominant on the battlefield, AFRL will invest in game-changing technologies and demonstrate their potential to provide new capabilities for the Air Force's global mission.

Objective 4.1: Foster basic research that probes beyond current technological limits.

AFRL must conduct mission-focused research that best serves the needs of the Air Force. The research must be balanced across the Core Function Support Plans, integrated across the TDs, and be on the cutting edge of research that leads to future Air Force capabilities. The Air Force Scientific Advisory Board (SAB) will periodically review AFRL's technical areas to assess whether AFRL is investing in relevant research across the near-, mid-, and far-terms.

Air Force Research Laboratory 2014 Strategic Plan

Objective 4.2: Lay the scientific and technical foundation for the future force.

AFRL will continue to conduct quality S&T by creating a knowledgeable workforce through organic in-house initiatives, partnering with the industrial base and academia, and through international relationships. These efforts help to build the technical breadth and depth necessary to advance our Air Force and preclude technological surprise by any potential adversary. The Air Force SAB will periodically review AFRL's quality by judging the rigor, equipment, facilities, and research conducted by AFRL.

Objective 4.3: Identify and demonstrate/simulate Game Changing Capabilities for Air Force leadership. Ensure the technical viability of the industrial base as appropriate.

AFRL will identify technology opportunities that provide "step-function" enhancements or revolutionary advancements in the lethality, survivability, affordability, and effectiveness of Air Force operations. AFRL will demonstrate the most promising Game-Changing technologies to speed their transition to the warfighter.

S&T Strategic Goal 5: Employ business practices that increase the S&T Enterprise's inventiveness, productivity, and responsiveness to Air Force needs.

Business practices are a key enabler for all other Strategic Goals. AFRL must take an inward look at business practices and policies to bolster or incentivize inventiveness, productivity, and responsiveness to Air Force needs. AFRL will seek to provide an environment conducive to creative thinking, innovation and problem solving, and will increase collaborations and partnerships to take advantage of advancement opportunities in other TDs, industry, and academia.

Objective 5.1: Improve and increase collaborations and partnerships with industry, academia, and other government organizations (include international).

AFRL can reduce the time and expense of developing new technologies, avoid technological surprise, and respond quickly to emerging threats and disruptive technologies by leveraging worldwide S&T investment. AFRL will investigate and use all the authorities and avenues available to optimize collaboration. AFRL will increase the use of small business to harness innovation and ingenuity required by our Air Force.

Objective 5.2: Identify, employ, and decentralize unique authorities to pilot process and productivity enhancements for AFRL.

To create an environment where inventiveness and innovation thrive, AFRL will identify and implement promising initiatives to increase the productivity of the science and engineering workforce and will decentralize authorities, when possible, to enable TDs to tailor processes to address the unique challenges and circumstances of each TD. The TDs will tackle its functional support processes and people in the same manner and spirit.

Objective 5.3: Reduce/streamline/eliminate outdated and ineffective policies and processes.

AFRL will examine existing business practices and will seek to improve or eliminate ineffective and outdated policies and processes.

Objective 5.4: Increase cross directorate planning and collaboration.

As Air Force capabilities become increasingly complex, it is important for AFRL to develop integrated solutions to produce technologies not otherwise achievable by one TD. As demonstrations of individual and integrated technologies become increasingly important, we will look to optimally leverage the skills, resources, and ideas found throughout the Laboratory to produce results that are only possible through partnering and integration.

S&T Strategic Goal 6: Acquire, develop, and retain a talented and high performing S&T workforce.

The success of AFRL depends on an agile, capable workforce that leads cutting-edge research, explores emerging technology areas, and promotes innovation across government, industry, and academia. The Air Force depends on the technical skill and exceptional aptitude of AFRL's science, engineering, and professional support workforce; in the defense industrial base; and in universities to successfully meet our national security objectives. AFRL must aggressively pursue strategic partnerships and outreach activities with schools, universities, sister services, professional associations, other Federal agencies, and international partners to grow the science, technology, engineering, and mathematics (STEM) workforce of the future. To acquire, develop, and retain such a workforce, AFRL will use the following objectives:

Objective 6.1: Pursue innovative methods to accelerate and improve the hiring process in order to hire highly talented personnel.

AFRL will investigate and implement innovative hiring processes within its control to attract and hire superior technical and professional talent to ensure an influx of talented and creative personnel. AFRL will also investigate hiring initiatives to reduce the administrative burden and speed up timelines; and will provide recommendations for new authorities designed to improve the hiring process.

Objective 6.2: Identify and implement the most effective workforce development initiatives.

Equally important to AFRL's future success is continuous development and retention of the current workforce. AFRL will investigate and employ workforce development initiatives that will ensure AFRL remains an attractive place to work and pursue a career to further AFRL's S&T mission.

Objective 6.3: Create STEM opportunities that inspire the future S&T workforce.

The competitive global research environment has created increased need for technically-capable personnel. In order to ensure the sufficient continuous availability of technically-competent personnel to conduct relevant research and sustain the S&T infrastructure needed by the Air Force, AFRL will continue to sponsor and seek out new opportunities for outreach initiatives focused on undergraduates, graduate students, and post-doctoral fellows that inspire them to pursue education and participation in STEM programs.

S&T Strategic Goal 7: Invest in core S&T infrastructure to ensure its future health.

Facilities and equipment are key enablers to effectively accomplish AFRL's mission and to attract quality people. AFRL will work to optimize the use of facilities and will ensure that it has access to the right mix of facilities needed to accomplish its mission, now and in the future. External agencies own and operate unique facilities that are frequently invaluable in furthering key S&T efforts. AFRL will incorporate these unique external facilities into its use plan for S&T facilities and equipment. To advance game-changing technologies, it is necessary for AFRL to forecast and resource new or improved facilities to support development of new capabilities. In an increasingly integrated digital world, AFRL must also increase IT capability across multiple networks, to ensure its scientists and engineers have the collaboration and research tools to effectively conduct its mission. Additionally, a viable industrial S&T base in the key technologies needed to meet Air Force needs is an essential component to the technology development and transition processes for application to weapon systems.

Objective 7.1: Optimize use of available research facilities across the AFRL Enterprise, leverage national S&T facilities, and divest, as appropriate.

AFRL will optimize the use of AFRL facilities and will divest from facilities that are no longer needed or are too costly to maintain given their utility, particularly if a comparable facility exists within the national infrastructure, is cost-effective to use, and can be counted on to be available when needed.

Objective 7.2: **Identify and resource new or improved facilities that enable the pursuit of game changing technologies.**

As new capabilities are pursued, AFRL must have access to state-of-the art research facilities. AFRL will look at future Air Force mission needs, identify the facilities required to develop the supporting technologies, and plan resources to build new or improved facilities.

Objective 7.3: **Extend the Defense Research and Engineering Network (DREN), Secret Internet Protocol Router Network (SIPRNet), and specialized ISR network access to provide a 21st Century IT infrastructure that supports effective collaboration and scientific research.**

Since IT cross-cuts the entire enterprise, AFRL will ensure the S&T workforce has a robust and resilient IT infrastructure and tools that enable effective collaboration and research. Key to this objective is wider implementation of and access to networks that provide capabilities not available on the Non-Secure Internet Protocol Router Network. Enabling our S&E workforce the ability to communicate and collaborate with industry, academia, and government entities virtually and in classified forums is critical to its future success.

Objective 7.4: **Foster the viability of the technical industrial base infrastructure, as appropriate.**

To maintain a strong Air Force, the technical industrial base must be strong as well. In keeping with national and DoD policy, AFRL will engage with industry to help keep key sectors of the United States' organic engineering and manufacturing industries healthy. AFRL will partner with the industrial base to identify then simulate and/or demonstrate game-changing capabilities along a spectrum from virtual constructs to open air ranges. AFRL plays a critical role in partnering with industrial sources of innovation and manufacturing capacity to ensure its responsiveness and relevancy and this must continue into the future.

AFRL Governance Structure

Through the AFRL corporate governance structure, metrics will be used to monitor progress towards meeting AFRL's Strategic Goals and Objectives. The Directors have the responsibility to apply the Strategic Plan's applicable goals and objectives within each organization and to ensure the goals and objectives are being met. The Plans and Programs Directorate has overall responsibility for tracking and reporting the status of goals and objectives in the Strategic Plan.

Conclusion

For many years, the Air Force S&T program through AFRL has provided leading edge S&T to advance the capabilities of the Air Force and will continue to do so far into the future. This Strategic Plan provides the AFRL workforce with an overarching guidance that is consistent with the Air Force S&T Strategy and other headquarters direction. The Strategic Goals describe our issues that are of critical importance to the Air Force. These goals should receive increased emphasis and serve as guideposts for resource decisions. By incorporating this guidance, AFRL will be a high-value contributor providing **_Revolutionary, Relevant and Responsive_** technologies that will ensure our Air Force will continue to be the best in the world.

AFB	Air Force Base
AFDD	Air Force Doctrine Document
AFMC	Air Force Materiel Command
AFOSR	Air Force Office of Scientific Research
AFRL	Air Force Research Laboratory
AGN&C	Aerodynamics, Guidance, Navigation & Control
C4I	Command, Control, Communications, Cyber, and Intelligence
DoD	Department of Defense
EO	Electro-Optical
HPW	Human Performance Wing
HSI	Human Systems Integration
IR	Infrared
ISR	Intelligence, Surveillance, and Reconnaissance
IT	Information Technology
PNT	Precision Navigation and Timing
RF	Radio Frequency
S&T	Science and Technology
SAB	Scientific Advisory Board
SOCCER	Senior Officer Communications and Coordination Electronic Resource
STEM	Science, Technology, Engineering, and Mathematics
TD	Technology Directorate
US	United States
USAF	United States Air Force

Air Force Research Laboratory 2014 Strategic Plan

Annex B: References

AFDD 1, Air Force Basic Doctrine, Organization, and Command, 14 October 2011

Air Combat Command Strategic Plan: Securing the High Ground 2012

Air Force Science & Technology Strategy

Air Force Strategic Environment Assessment, 2012-2032, 11 March 2011

Air Force Materiel Command (AFMC) 2013 Strategic Plan, January 2013

A National Strategic Plan for Advanced Manufacturing, National Science and Technology Council, February 2012

Bright Horizons the Air Force Science, Technology, Engineering and Math (STEM) Workforce Strategic Roadmap, 16 March 2011

Defense Budget Priorities and Choices, Fiscal Year 2014, April 2013

Global Horizons Final Report: United States Air Force Global Science and Technology Vision, 21 June 2013

Science and Engineering Indicators, National Science Board, February 2012

Science and Technology Priorities for the FY 2014 Budget, Office of Management and Budget, 6 June 2012

Sustaining US Global Leadership: Priorities for a 21st Century Defense, January 2012

Technology Horizons: A Vision for Air Force Science & Technology during 2010-2030, 15 May 2010

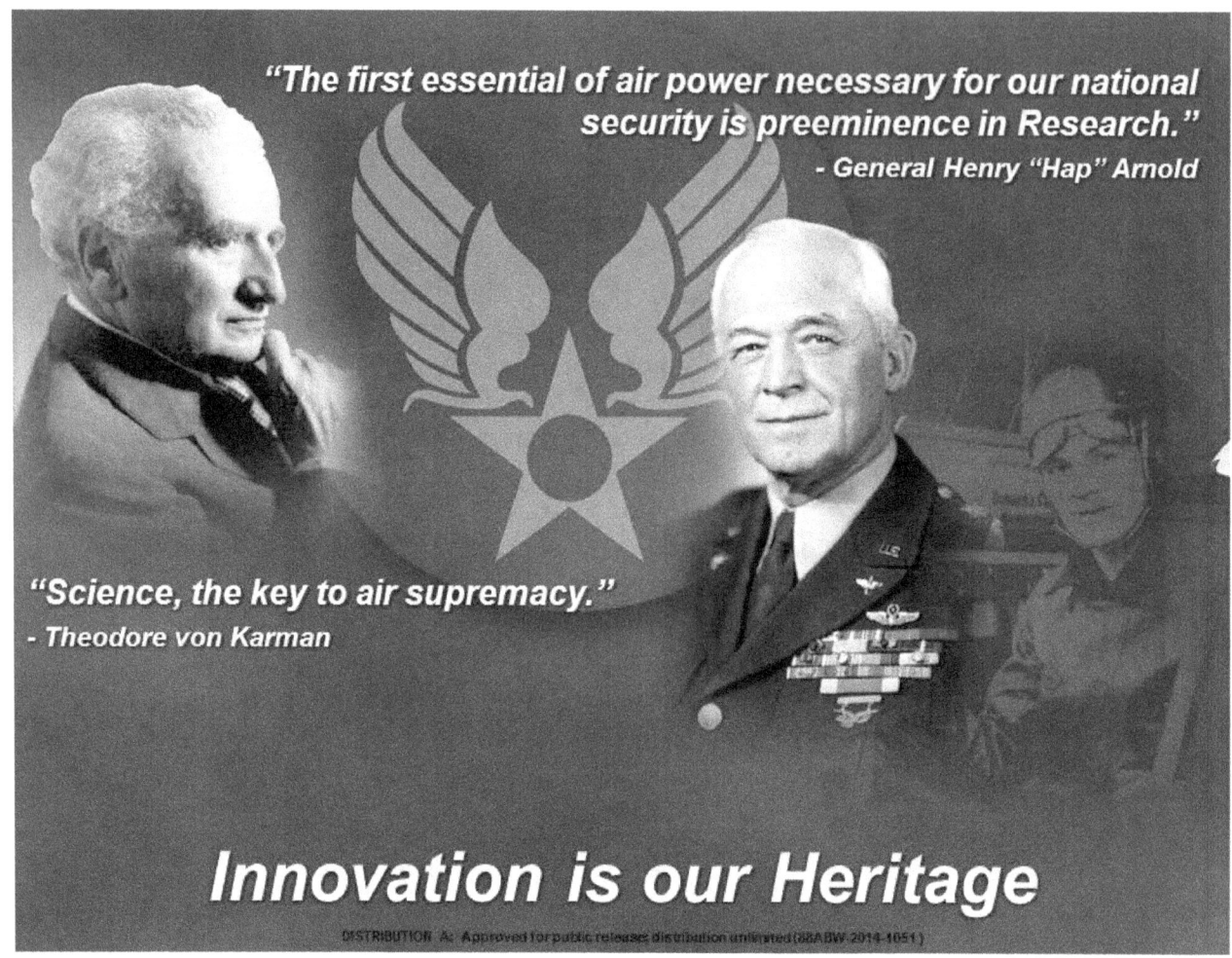